ASCE STANDARD

ANSI/ASCE/EWRI

65-17

Calculation of the Saturated Hydraulic Conductivity of Fine-Grained Soils

ASCE AMERICAN SOCIETY OF CIVIL ENGINEERS

EWRI ASCE

Approved American National Standard ANSI

PUBLISHED BY THE AMERICAN SOCIETY OF CIVIL ENGINEERS

Library of Congress Cataloging-in-Publication Data

Names: American Society of Civil Engineers.
Title: Calculation of the saturated hydraulic conductivity of fine-grained soils: ASCE STANDARD, ANSI/ASCE/EWRI 65-17.
Other titles: ASCE STANDARD, ANSI/ASCE/EWRI 65-17
Description: Reston, Virginia : American Society of Civil Engineers, [2017] | Includes bibliographical references and index.
Identifiers: LCCN 2017003718 | ISBN 9780784414347 (soft cover) | ISBN 9780784480496 (PDF)
Subjects: LCSH: Soils–Testing–Standards–United States. | Soil permeability–Measurement–Standards–United States. | Groundwater flow–Measurement–Standards–United States.
Classification: LCC TA710.5 .C345 2017 | DDC 624.1/5136–dc23 LC record available at https://lccn.loc.gov/2017003718

Published by American Society of Civil Engineers
1801 Alexander Bell Drive
Reston, Virginia, 20191-4382
www.asce.org/bookstore | ascelibrary.org

This standard was developed by a consensus standards development process which has been accredited by the American National Standards Institute (ANSI). Accreditation by ANSI, a voluntary accreditation body representing public and private sector standards development organizations in the United States and abroad, signifies that the standards development process used by ASCE has met the ANSI requirements for openness, balance, consensus, and due process.

While ASCE's process is designed to promote standards that reflect a fair and reasoned consensus among all interested participants, while preserving the public health, safety, and welfare that is paramount to its mission, it has not made an independent assessment of and does not warrant the accuracy, completeness, suitability, or utility of any information, apparatus, product, or process discussed herein. ASCE does not intend, nor should anyone interpret, ASCE's standards to replace the sound judgment of a competent professional, having knowledge and experience in the appropriate field(s) of practice, nor to substitute for the standard of care required of such professionals in interpreting and applying the contents of this standard.

ASCE has no authority to enforce compliance with its standards and does not undertake to certify products for compliance or to render any professional services to any person or entity.

Errata: Errata, if any, can be found at http://dx.doi.org/10.1061/9780784414347.

ISBN 978-0-7844-1434-7 (print)
ISBN 978-0-7844-8049-6 (PDF)
Manufactured in the United States of America.

24 23 22 21 20 19 18 17 1 2 3 4 5

ASCE STANDARDS

In 2014, the Board of Direction approved revisions to the ASCE Rules for Standards Committees to govern the writing and maintenance of standards developed by ASCE. All such standards are developed by a consensus standards process managed by the ASCE Codes and Standards Committee (CSC). The consensus process includes balloting by a balanced standards committee, and reviewing during a public comment period. All standards are updated or reaffirmed by the same process every 5 to 10 years. Requests for formal interpretations shall be processed in accordance with Section 7 of ASCE Rules for Standards Committees which are available at www.asce.org. Errata, addenda, supplements, and interpretations, if any, for this standard can also be found at www.asce.org.

Standard ANSI/ASCE/EWRI 65-17, *Calculation of the Saturated Hydraulic Conductivity of Fine-Grained Soils* was approved by the Codes and Standards Committee with a waiver for the units in the standard. The waiver is for only International System of Units (SI) to be used in the standard.

This standard has been prepared in accordance with recognized engineering principles and should not be used without the user's competent knowledge for a given application. The publication of this standard by ASCE is not intended to warrant that the information contained therein is suitable for any general or specific use, and ASCE takes no position respecting the validity of patent rights. The user is advised that the determination of patent rights or risk of infringement is entirely their own responsibility.

A complete list of currently available standards is available in the ASCE Library (http://ascelibrary.org/page/books/s-standards).

CONTENTS

PREFACE

This is a standard guideline for calculating saturated hydraulic conductivity (K_{sat}), permeability (k), and porosity (n) of fine-grained soils using (1) strain-stress data from the incremental loading of a soil sample in a standardized consolidometer (step-load test), (2) one-dimensional vertical consolidation theory relating K_{sat} to the coefficient of consolidation (c_v), (3) the relation between K_{sat} and k, and (4) the relation between the porosity and the void ratio of a soil undergoing primary consolidation. An undisturbed soil specimen is inserted in a consolidometer and subjected to incremental vertical loads allowing axial (vertical) drainage. The vertical axial deformation of the soil specimen is measured as a function of elapsed testing time through various cycles of incremental loading until the end of the consolidation test. The measured deformation is graphed as a function of the $\log_{10}$ of time or the square root of time for each increment of vertical load. These graphs are used to calculate the coefficient of consolidation for each loading increment applied to the soil specimen. Terzaghi's (1925) one-dimensional (1D) vertical consolidation theory relating K_{sat} to c_v is then applied to calculate K_{sat} for each increment of vertical load applied to the soil specimen. The permeability and porosity of the soil sample decrease with increasing applied vertical load according to known laws. The permeability and porosity of the tested soil specimen are calculated for each load increment as well. The calculated K_{sat}, k, and porosity yield a series of pairs of values (K_{sat}, vertical effective stress; k, vertical effective stress; and porosity, vertical effective stress) that are graphed to produce relations between (1) K_{sat} and vertical effective stress, (2) permeability and vertical effective stress, and (3) porosity and vertical effective stress. This standard guideline demonstrates the relation between 1D consolidation and land subsidence driven by groundwater withdrawal, highlighting the relevance of strain-stress phenomena observed in consolidometer testing to the groundwater-induced phenomenon of land subsidence (consolidation of compressible strata by groundwater withdrawal). The increase in vertical effective stress caused by declining pore water pressure during sustained groundwater extraction; the changes effected by groundwater withdrawal on K_{sat}, k, and porosity; and the associated land subsidence are the primary foci of this standard guideline. This standard's methodology can be applied to refine or improve calculations of land subsidence, groundwater flow predictions, and transport of dissolved solutes moving in groundwater through fine-grained soils. The methods of this standard guideline are limited to fine-grained, compressible, inorganic soils. Highly organic soils (peat) exhibit anomalous primary and secondary consolidation not within the scope of this standard guideline.

This standard guideline represents the consensus of the Standards Committee on Fitting of Hydraulic Conductivity using Statistical Spatial Estimation (called KSTAT) of the Standards Development Council (SDC) of the Environmental and Water Resources Institute (EWRI) of ASCE. This standard guideline is the fourth in a series of standards that seeks to enhance the probabilistic and empirical characterization and understanding of the saturated hydraulic conductivity (K_{sat}), a key groundwater parameter. The KSTAT Standards Committee has published three companion standard guidelines: Standard ASCE/EWRI 50-08 (ASCE 2008a), Standard ASCE/EWRI 51-08 (ASCE 2008b), and Standard ASCE/EWRI 54-10 (ASCE 2010). Standard ASCE/EWRI 50-08 addresses the optimal fitting of saturated hydraulic conductivity (K_{sat}) with skewed probability density functions (pdfs). Standard ASCE/EWRI 51-08 deals with the estimation of the effective saturated hydraulic conductivity, a parameter that relates the average specific discharge to the average hydraulic gradient. Standard ASCE/EWRI 54-10 presents a methodology for the geostatistical interpolation and block averaging of K_{sat} in statistically homogeneous and isotropic aquifers.

The formulas in this standard guideline require that all their values be expressed in the same system of units, be it the International System of Units (SI) (meter, second, Newton, etc.) or the common system of units in the United States (feet, second, pound, etc.).

The provisions of this document are written in permissive language and, as such, offer to the user a series of options or instructions but do not prescribe a specific course of action. Significant judgment is left to the user of this document.

ASCE does not endorse commercial spreadsheets, numerical software, or testing methods produced by other organizations cited in this standard guideline. Any such products are cited in this standard guideline to illustrate possible ways of carrying out calculations and conducting experimental tests. It is left to the user's discretion to choose and verify the accuracy of whichever computational technique or testing method to apply to implement the methodology.

ACKNOWLEDGMENTS

The American Society of Civil Engineers and the Environmental and Water Resources Institute acknowledge the work of the Standards Committee on Fitting of Hydraulic Conductivity Using Statistical Spatial Estimation (called KSTAT). The committee members are professionals from academic, governmental, and private sectors. The current members of the Standards Committee on Fitting of Hydraulic Conductivity Using Statistical Spatial Estimation are

Nazeer Ahmed, Ph.D., P.E., M.ASCE
Teresa Culver, Ph.D, M.ASCE
Macan Doroudian, Ph.D, P.E., G.E., M.ASCE
Mohamed M. Hantush, Ph.D., M.ASCE
Paul F. Hudak, Ph.D.
Hugo A. Loáiciga (Chairman), Ph.D., P.E., F.ASCE
Miguel A. Marino, Ph.D., Dist.M.ASCE
Laurent M. Meillier, P.G.
Chin Man W. Mok, Ph.D., P.E., P.G., G.E., D.WRE, D.GE, F.ASCE, F.EWRI
Kok-Kwang Phoon, Ph.D., P.E., F.ASCE
George F. Pinder, Ph.D., Dist.M.ASCE
Anand J. Puppala, Ph.D., P.E., F.ASCE
Mark Henry Rubarenzya, Ph.D., D.WRE, M.ASCE
Zhuping Sheng, Ph.D. P.E., P.H., F.ASCE
Parmeshwar L. Shrestha, Ph.D., P.E., D.WRE, M.ASCE
Stewart W. Taylor, Ph.D., P.E., F.ASCE
Frank T.-C. Tsai, Ph.D., P.E., M.ASCE
William W.-G. Yeh, Ph.D., Dist.M.ASCE

CHAPTER 1
SCOPE

This is a standard guideline for calculating the saturated hydraulic conductivity (K_{sat}), permeability (k), and porosity (n) of fine-grained, isotropic, and homogeneous soils using (1) strain-stress data from the incremental loading of a soil sample in a standardized consolidometer (step-load test), (2) 1D vertical consolidation theory relating K_{sat} to the coefficient of consolidation (c_v), (3) the relation between K_{sat} and k, and (4) the relation between porosity and the void ratio of a soil undergoing primary consolidation. An undisturbed soil specimen is inserted in a consolidometer and subjected to incremental vertical loads allowing axial (vertical) drainage. The vertical axial deformation of the soil specimen is measured as a function of elapsed testing time through several cycles of incremental loading until the end of the consolidation test. The measured deformation is graphed as a function of the $\log_{10}$ of time or the square root of time for each increment of vertical load. These graphs are used to calculate the coefficient of consolidation for each load increment applied to the soil specimen. (Casagrande's unpublished work of 1938 and Taylor's 1948 graphical methods are commonly used.) Terzaghi's (1925) 1D vertical consolidation theory relating K_{sat} to c_v is then used to calculate K_{sat} for each increment of vertical load applied to the soil specimen. The permeability and porosity of the soil sample decrease with increasing applied vertical load according to known laws. The permeability and porosity of the tested soil specimen are calculated for each loading increment as well. The calculated K_{sat}, k, and porosity yield a series of pairs of values (K_{sat}, vertical effective stress; k, vertical effective stress; and porosity, vertical effective stress) that are graphed to produce relations between (1) K_{sat} and the vertical effective stress, (2) permeability and vertical effective stress, and (3) porosity and vertical effective stress.

The classical theory of 1D (vertical) consolidation (Terzaghi 1925) assumes that K_{sat}, k, and porosity remain constant as soils consolidate under increasing vertical effective stress caused by external loads or groundwater extraction. In fact, the poroelastic theory of 3D consolidation makes the same assumption concerning K_{sat} of a deforming soil (Loáiciga 2013). Yet, the experimental evidence from 1D consolidation tests indicates that K_{sat}, k, and porosity decrease when a soil consolidates under increasing vertical effective stress as discussed in this document. The increase in vertical effective stress caused by declining pore water pressure during sustained groundwater extraction; the changes effected by groundwater withdrawal on K_{sat}, k, and porosity; and the associated land subsidence are the primary foci of this standard guideline. This standard guideline demonstrates the relation between 1D consolidation and land subsidence driven by groundwater withdrawal, highlighting the relevance of strain-stress phenomena observed in consolidometer testing to the groundwater-induced phenomenon of land subsidence (consolidation of compressible strata by groundwater withdrawal). This standard's methodology can be applied to calculations of land subsidence, groundwater flow predictions, and transport of dissolved solutes moving in groundwater through fine-grained soils.

CHAPTER 2
TERMINOLOGY

Coefficient of permeability—The laminar flow of groundwater per unit area of soil perpendicular to the direction of flow under a unit hydraulic gradient and standard water temperature [commonly 20°C (ASTM 2011a)]. This term is used in the geotechnical engineering and engineering geology communities. See Saturated hydraulic conductivity.

Consolidation—The reduction of the volume of a soil mass caused by the increase of effective stress.

Consolidation settlement—The drop in elevation of a soil surface subjected to an increase of effective stress or the reduction in height of a soil specimen compressed in a consolidation test.

Effective stress—The average, normal force per unit area transmitted from grain to grain of a soil mass. In most applications of geotechnical engineering it is calculated as the total stress minus the pore pressure at a given location within a soil or rock mass. See Nur and Byerlee (1971) for an exact calculation formula for effective stress. If the normal force is vertical (or horizontal), the resulting stress is called vertical (or horizontal) effective stress.

Fine-grained soil—A soil with 50% or more of its dry mass passing the U.S. standard sieve #200. Fine-grained soils have relatively low saturated hydraulic conductivity, commonly less than 1×10^{-3}cm/s.

Hydraulic gradient—The change of hydraulic head per unit length of motion of water in a soil mass.

Hydraulic head—The potential energy per unit weight of pore water (elevation head) plus the mechanical energy per unit weight of pore water (pressure head) at a given location within a soil mass. It has units of length.

Intrinsic permeability—See Permeability.

Land subsidence—The drop of elevation of the land surface in an area underlain by aquifers from which groundwater is withdrawn.

Permeability—A geometric property of the pores of a soil mass that measures the capacity of fluids to move through the soil mass driven by a potential gradient. It has units of length squared and is commonly expressed in darcys (1 darcy $= 9.87 \times 10^{-9}$cm^2), millidarcys, or microdarcys. It is sometimes called intrinsic permeability to avoid confusing it with the coefficient of permeability, the latter used by some authors to describe what this standard guideline defines as the saturated hydraulic conductivity.

Poroelastic aquifer—Porous medium that is assumed to have the properties of an elastic solid and in addition can store and transmit water.

Porosity—The ratio of the volume of pore to the total volume of a soil or rock specimen.

Primary consolidation settlement—The drop in elevation of a soil surface subjected to an increase of compressive stress after all the excess pore pressure has dissipated or the reduction in height of a soil specimen subjected to an increase of compressive stress in a consolidation test after all the excess pore pressure has dissipated.

Saturated hydraulic conductivity—The laminar flow of groundwater per unit area of soil perpendicular to the direction of flow under a unit hydraulic gradient and standard water temperature [commonly 20°C (ASTM 2011a)]. This term, or simply hydraulic conductivity, is standard in the groundwater hydrology and hydrogeology communities. See Coefficient of permeability.

Total stress—The normal force per unit area of soil. If the normal stress is vertical (or horizontal), it is called vertical (or horizontal) total stress.

Undisturbed specimen (soil or soil sample)—Sampled mass of soil that has identical composition, properties, and conditions as the soil in situ. In reality, it is nearly impossible to obtain undisturbed samples of soil owing to the effects of drilling or excavation, sample retrieval, removal of confinement, sample storage and transportation, and laboratory processing (Marcuson and Franklin 1979). For this reason, it is practical to obtain samples of intact soil (and rock) specimens following the state of the practice or standard of care to preserve in situ structure, water content, density, and other properties to a level consistent with the intended purpose for testing (ASTM 2011a). Groups C and D soil samples in the ASTM terminology are physically intact. Their collection frequently involves the use of thin-wall or double-rotating core sampling devices (ASTM 2005).

Vertical (or horizontal) effective stress—See Effective stress.

Vertical (or horizontal) total stress—See Total stress.

Void ratio—The ratio of the volume of pores to the volume of solids in a soil specimen.

CHAPTER 3
SUMMARY OF THE CONSOLIDATION TEST

3.1 INCREMENTAL LOADING

Fig. 3-1 shows the main features of a general-purpose consolidometer. Details about standardized consolidometers and the step-load test (incremental vertical loading of a soil specimen inside a loading device) are found in the American Society of Testing and Materials (ASTM) test method D2435-08 (ASTM 2011b) and in test method EM1119-2-1906 of the U.S. Army Corps of Engineers (USACE 1980). For completeness, a brief summary of the step-load test is included here.

An undisturbed soil specimen (or sample) of fine-grained soil (see Chapter 2, Terminology) is prepared and inserted into a specimen ring. The soil specimen and ring are placed in the loading device. The soil specimen's diameter-to-height ratio ranges between 2.5 and 5. The ring's rigidity is such that it prevents lateral deformation of the soil specimen, thus producing vertical (1D) strain. An initial linear variable differential transducer (LVDT) reading is taken of the height of the soil specimen once it is inside the loading device. Typical measurement precision is $\pm 10^{-4}$ in. ($\cong \pm 2$ μm). The loading plate and top porous stone are seated onto the specimen exerting a small seating pressure. A first increment of load (P) and normal stress (P/A) is applied to the soil specimen, where A is the soil specimen's cross-sectional area. The cross-sectional area of the soil specimen remains constant during consolidation because of the rigidity of the ring confining the soil specimen. The first stress level typically lies in the range 10 to 25 kPa ($1\ \text{kPa} = 1\ \text{kN/m}^2$). The soil specimen is inundated to induce full saturation. Saturation may require applying a back-pressure to the soil specimen. The loading may have to be increased to prevent swelling if the soil is expansive until all swelling ceases. The soil specimen consolidates under the initial load P as vertical drainage takes place through the upper and lower porous plates (implying double drainage in the soil specimen, that is, upward and downward vertical drainage) until the deformation becomes negligible. The loading time for each loading increment is commonly limited to 24 h for extremely slow-draining soil specimens. The deformation of the soil specimen is measured over time with an LVDT.

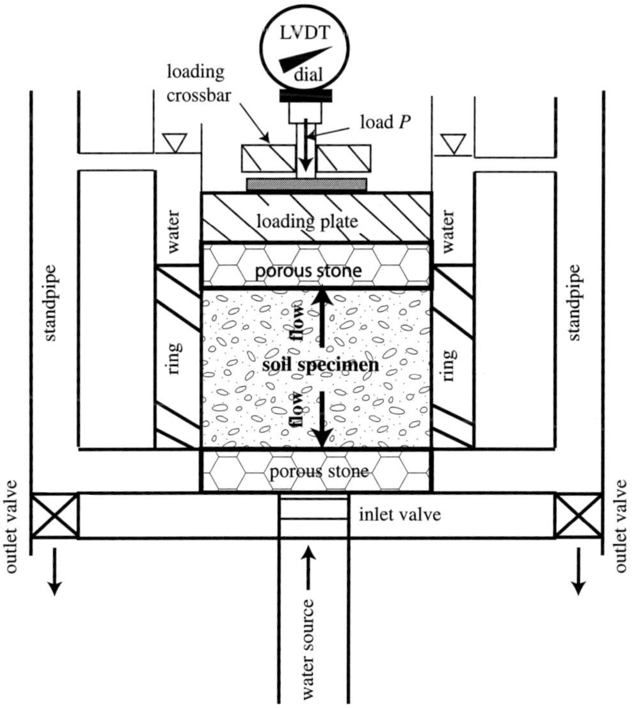

FIGURE 3-1. Generic Features of a Consolidometer. LDVT: Linear Variable Differential Transducer, Measures Deformation of the Soil Specimen

In the second step of the consolidation test the load is increased to twice the magnitude of the initial load P, and the sample is allowed to complete a new cycle of consolidation under the load $2P$ (normal stress equals $2P/A$) while deformation is measured. This process of reloading the sample with a new load twice the magnitude of the previous load and measuring deformations is repeated until the maximum load is applied, marking the end of the 1D consolidation test. It is common during consolidation tests to reduce the load several times once a level of loading is achieved to measure the unloading response of the soil specimen. Thereafter, the load is increased again to complete an unloading-reloading loop and continue with the consolidation test until reaching the maximum test load. Fig. 3-2 illustrates the incremental loading approach of the consolidation test.

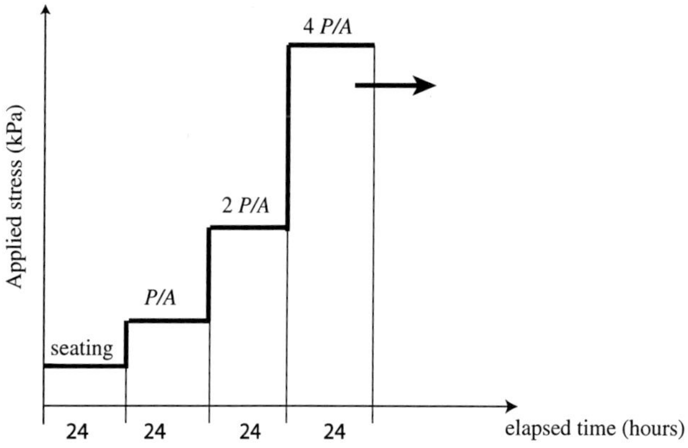

FIGURE 3-2. Step-Loading Method with Repetitive Doubling of the Applied Vertical Stress

3.2 DEFORMATION, STRAIN, VOID RATIO, AND COEFFICIENT OF COMPRESSIBILITY

Fig. 3-3 depicts the deformation of the soil specimen inside a consolidometer. In Fig. 3-3, the height of the soil specimen prior to the application of the load P (change in vertical effective stress $\Delta\sigma_v' = P/A$) equals H_1. At the end of the first load increment, the height of the specimen equals $H_1 - \Delta H_1$, which is also the height of the specimen at the beginning of the second loading increment under load $2P$ (change in vertical effective stress $\Delta\sigma_v' = 2P/A$). At the end of the second loading increment, the soil specimen's height stands at $H_1 - \Delta H_2$. The height of the soil specimen continues to be reduced with repeated incremental loading until reaching a final height, H_f, at the end of the consolidation test. The LDVT measures the deformation (ΔH) of the specimen over time during each loading increment for all loading increments of the consolidation test. The height of solids, H_s (or volume of solids per unit cross-sectional area of soil specimen), is assumed to remain unaltered through the consolidation test.

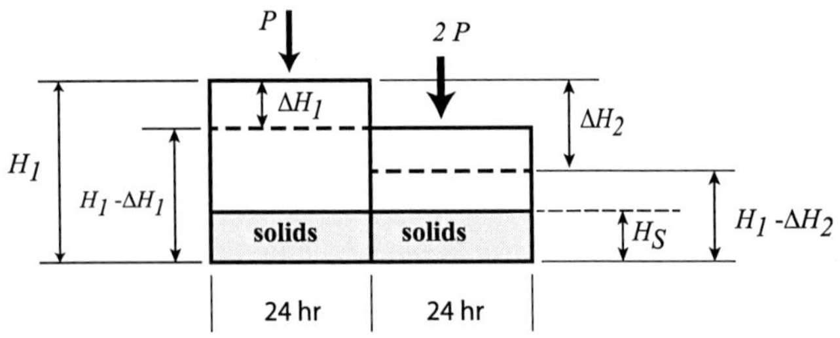

FIGURE 3-3. Change of the Pore Volume and Specimen Height During the Application of the Incremental Loads *P* and 2*P*

Let e_0 and H_0 be the initial void ratio and height of a soil specimen placed inside the consolidometer, and R_0 denote the initial reading in the LVDT measuring the deformation of the soil specimen. The reading in the LDVT at elapsed time, t, after incremental loading begins is R_t, and the deformation, height, strain, and void ratio of the soil specimen at time t are denoted by ΔH_t, H_t, ε_t, and e_t, respectively. The following formulas comprise the various relations among these consolidation variables. (The initial strain ε_0 is considered equal to zero in 3-3.)

$$\Delta H_t = H_t - H_0 = R_t - R_0 \tag{3-1}$$

$$\varepsilon_t = \frac{\Delta H_t}{H_0} \tag{3-2}$$

$$e_t = e_0 - \varepsilon_t \cdot (1 + e_0) \tag{3-3}$$

Another deformation-related variable is the coefficient of compressibility (a_v), which is the change of void ratio per unit change in vertical effective stress:

$$a_v = \frac{e_0 - e_t}{\sigma_{vt}' - \sigma_{v0}'} \tag{3-4}$$

in which σ_{vt}' and σ_{v0}' are the applied vertical effective stresses corresponding to the void ratios e_t and e_0, respectively.

CHAPTER 4
CHANGES IN STRESS AND DEFORMATION OF STRATA

Fig. 4-1 shows a typical situation in the field whereby a compressible soil stratum is loaded by the construction of an embankment. The stratum is subjected to a vertical stress, q, at the base of the embankment. The stress q equals the weight of the embankment (W) divided by its basal area (A). A soil element, X, at depth z within the stratum is subjected to vertical and horizontal effective stresses σ'_{v0} and σ'_{h0}, respectively, under an unloaded condition. The element sustains an increase in vertical effective stress $\Delta\sigma'_v$ and in lateral effective stress $\Delta\sigma'_h$ (in the plane of Fig. 4-1 and perpendicular to it) and deforms as shown in Fig. 4-1. The changes in vertical and lateral effective stresses are commonly approximated with results from the theory of elasticity (Boussinesq's or Westergaard's solutions are commonly used), approximated using empirical methods (see Holtz et al. 2011 for details about stresses induced by vertical loads), or by numerical simulation of stress-strain in the loaded soils (Griffiths and Marquez 2007). The deformation of the soil element X is 3D in general, because the basal vertical stress q induces changes of stress in all directions in the soil stratum. The soil element X would undergo approximate 1D vertical deformation if the basal dimensions of the embankment were very large compared to the thickness (H_0) of the stratum (Skempton and Bjerrum 1957). One-dimensional vertical deformation is achieved in the 1D consolidation test by restricting lateral deformation of the soil specimen.

Fig. 4-2 illustrates the changes in the stress field of a soil element X found at a depth z inside a compressible soil stratum of initial thickness H_0 undergoing groundwater withdrawal at a rate Q. The soil element is a distance r from a well. Two cones of depression are shown in Fig. 4-2 depicting the progressively lower position of the phreatic surface (or "water table") at increasingly longer elapsed times since the beginning of groundwater withdrawal. Groundwater withdrawal raises the vertical (and lateral) effective stress acting on the element X by reducing the pore pressure $u(z)$, which acts isotropically at any depth z in the soil stratum. There is no application of external vertical loads, as is the case with, for example, the construction of embankments as portrayed in Fig. 4-1.

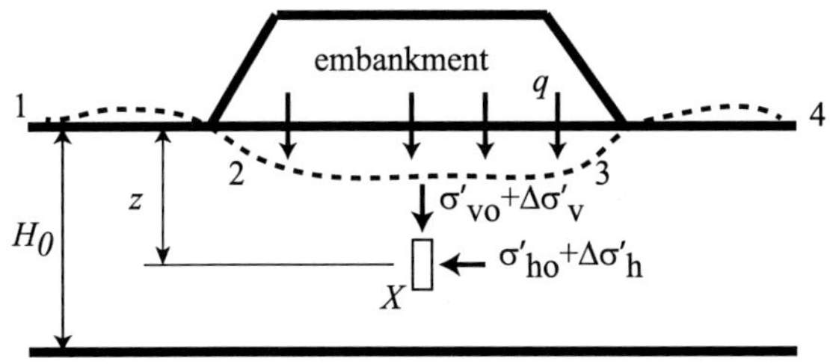

FIGURE 4-1. Deformation of a Compressible Stratum and a Soil Element by Embankment Loading

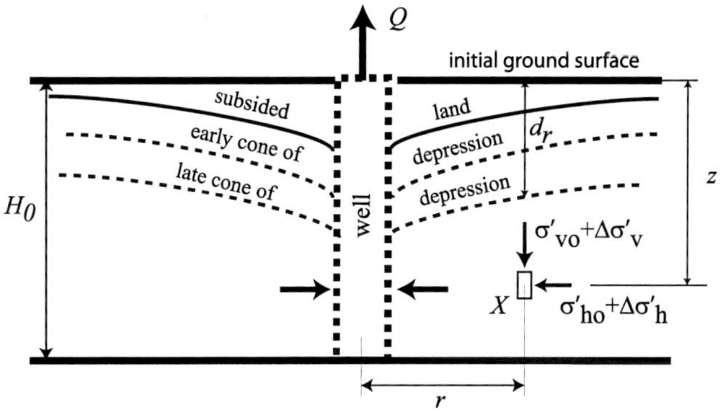

FIGURE 4-2. Deformation of a Compressible Stratum and Soil Element by Groundwater Extraction

Assume that the initial position of the phreatic surface in the stratum graphed in Fig. 4-2 coincides with the ground surface and that the saturated unit weight of the soil is γ. Let the unit weight of groundwater be $\gamma_w \cong 9.8\,\text{kN/m}^3$ and K_0 denote the at-rest coefficient of lateral earth pressure, which equals the ratio of the initial horizontal (lateral) effective stress (σ'_{ho}) to the initial vertical effective stress (σ'_{vo}) acting on the soil element X, shown in Fig. 4-2. The initial effective stresses, prior to groundwater withdrawal, acting on the soil element X are

$$\sigma'_{v0}(z) = \gamma \cdot z - \gamma_w \cdot z \tag{4-1}$$

$$\sigma'_{h0}(z) = K_0 \cdot \sigma'_{v0} \tag{4-2}$$

Suppose that groundwater withdrawal produces a steady-state drawdown d_r at a distance r from the extraction well and that due to capillarity the unit weight of the soil above and below the phreatic surface does not change appreciably. The effective stresses acting on the element X change to (neglecting vertical gradients of hydraulic head)

$$\sigma'_v(z,r) = \gamma \cdot z - \gamma_w \cdot (z - d_r) \tag{4-3}$$

$$\sigma'_h(z,r) = K_0 \cdot \sigma'_v \tag{4-4}$$

The increase in vertical effective stress would be

$$\Delta\sigma'_v(z,r) = \sigma'_v(z,r) - \sigma'_{v0}(z) = \gamma_w \cdot d_r \tag{4-5}$$

whereas the change in horizontal effective stress would equal (assuming that the at-rest coefficient remains unchanged)

$$\Delta\sigma'_h(z,r) = K_0 \cdot \gamma_w \cdot d_r \qquad (4\text{-}6)$$

The drawdown d_r is calculated from solution of the groundwater flow equation for radial flow (see Loáiciga 2013).

The increases in vertical and horizontal effective stress represented by Eqs. (4-5) and (4-6) are caused by the change in pore pressure alone. Their calculation does not involve special theories, as is the case with the changes in effective stress caused by the application of vertical loads, such as the embankment shown in Fig. 4-1. If the compressible stratum is isotropic, linear, and elastic, the vertical strain associated with the changes in vertical and horizontal effective stresses at a distance r from a well and at a depth z given by Eqs. (4-5) and (4-6) are derived from the 3D theory of elasticity (see Timoshenko and Goodier 1970 for a review of the theory of elasticity, and Ingebritsen et al. 2006 for an exposition of stress-strain in porous media) to yield

$$\varepsilon_v(z,\ r) = \frac{\gamma_w d_r}{E}(1 - 2\nu K_0)\ \ d_r \le z \le H_0 \qquad (4\text{-}7)$$

in which E and ν denote the elastic (or Young's) modulus and Poisson's ratio of the soil, respectively. The vertical strain for soil elements above the phreatic surface is (neglecting capillary pressures above the phreatic surface)

$$\varepsilon_v(z,\ r) = \frac{\gamma_w z}{E}(1 - 2\nu K_0)\ \ 0 \le z < d_r \qquad (4\text{-}8)$$

An integration of the strain in Eqs. (4-7) and (4-8) over the thickness H_0 of the stratum and summation of the integration results gives the land subsidence (or consolidation settlement) δ_r at a distance r from the well:

$$\delta_r = \frac{\gamma_w d_r}{E}(1 - 2\nu K_0) \cdot \left(H_0 - \frac{d_r}{2}\right) \qquad (4\text{-}9)$$

For example, with $d_r = 30$ m, $H_0 = 100$ m, and common values of elastic properties being $\nu = 0.4$, $K_0 = 0.5$ (normally consolidated soil) and $E = 50$ MPa (see Bowles 1996 for typical values of soil characteristics), the consolidation settlement would be $\delta_r = 0.30$ m at a distance r from an extraction well. The approach previously outlined is applicable to compressible aquitards imbedded in permeable formations. Some authors have approached the simulation of land subsidence solving numerically the equations of 3D deformation of a poroelastic aquifer (Gambolati et al. 2000).

Soils do not behave elastically except at very small strains, and in the case of highly organic soils perhaps not even then (Duncan and Chang 1970). Nevertheless, Eq. (4-9) reflects key correct trends in soil deformation. Specifically, it shows that the consolidation settlement increases with increasing drawdown (which, in turn, increases with increasing rate of groundwater withdrawal), it decreases with increasing soil stiffness (represented by the elastic modulus), and it decreases with increasing Poisson's ratio. Several approaches have been proposed to cope with the nonlinear nature of strain-stress relationships in soils in simulating consolidation settlement and land subsidence (Duncan and Chang 1970). A common approach is the assumption of 1D subsidence coupled with numerical groundwater flow simulation (the MODFLOW model with the 1D subsidence package SUB is an example; see also Loáiciga 2013, for another experimental-numerically based method). Another approach specifies nonlinear stress-strain functions and couples those with numerical groundwater simulation in 1D or 3D (Kang and Li 2015). An approach for the simulation of soil deformation under a variety of stresses is by means of the numerical solution of nonlinear elastoplastic finite-element equations of a deformable soil mass (Griffiths and Marquez 2007).

The effect of lowering of the phreatic surface extends tens of meters to hundreds of meters from an extraction well, depending on the rate of groundwater withdrawal and the hydraulic characteristics of the soil. The amount of subsidence decreases with increasing distance from extraction wells over large areas affected by groundwater withdrawal (Zektser et al. 2005). Groundwater extraction causes land subsidence that is predominantly 1D vertical settlement and induces changes in the saturated hydraulic conductivity, permeability, and porosity in a dewatered stratum in a manner analogous to the changes that occur under vertical loading applied to a soil specimen in a consolidometer. This analogy serves as a foundation for the methodology of this standard guideline.

CHAPTER 5
VARIABLE PROPERTIES OF FINE-GRAINED SOILS

Terzaghi's (1925) theory for 1D vertical consolidation of saturated soil subjected to vertical loading proposed the following groundwater flow equation in terms of the excess pore pressure (u_e) (pressure in excess of that prevailing prior to the application of loads to a soil) created by the application of loads:

$$c_v \frac{\partial^2 u_e}{\partial z^2} = \frac{\partial u_e}{\partial t} \quad (5\text{-}1)$$

in which c_v denotes the coefficient of consolidation. In Eq. (5-1), two boundary conditions specify zero excess pore pressure on the top and bottom permeable boundaries of a soil stratum in the case of double drainage or zero excess pressure on the permeable boundary and zero flow on the impermeable one in the case of single drainage. There is, in addition, an initial condition (at time zero) that specifies the distribution of the excess pore pressure throughout the soil stratum at the start of loading. The initial condition depends on the nature of the applied loads (Das 2015). The coefficient of consolidation depends on a number of basic properties: saturated hydraulic conductivity (K_{sat}), unit weight of water (γ_w), void ratio of the soil at the start of a loading increment (e_0), and the coefficient of compressibility (a_v), as follows:

$$c_v = \frac{K_{sat}}{\gamma_w} \cdot \frac{1+e_0}{a_v} \quad (5\text{-}2)$$

The consolidation coefficient (c_v) and the coefficient of compression (a_v) are determined by consolidometer testing. Formula (5-2) is inverted to calculate the saturated hydraulic conductivity (K_{sat}, dimensions of length over time):

$$K_{sat} = \frac{c_v \gamma_w a_v}{1+e_0} \quad (5\text{-}3)$$

The permeability (k, dimensions of length squared) is related to the saturated hydraulic conductivity by the following formula:

$$K_{sat} = k\frac{\gamma_w}{\mu} \quad (5\text{-}4)$$

in which μ represents the dynamic viscosity of groundwater, a function of groundwater temperature and density (see Fetter 2001). The permeability is sometimes referred to as intrinsic permeability to avoid confusion stemming from the word permeability being synonymous with hydraulic conductivity. Whereas the saturated hydraulic conductivity is a function of the characteristics of water (it depends on water viscosity and unit weight), the permeability, conversely, is a geometric property of a soil. It does not depend on the characteristics of the groundwater flowing through it. This permits converting a saturated hydraulic conductivity measured at a temperature, T [herein denoted by $K_{sat}(T)$], to a saturated hydraulic conductivity at a reference temperature, T_0 [herein denoted by $K_{sat}(T_0)$], as follows:

$$K_{sat}(T_0) = K_{sat}(T)\frac{\gamma_{wT_0}}{\mu_{T_0}}\frac{\mu_T}{\gamma_{wT}} \quad (5\text{-}5)$$

in which (μ_{T_0}, γ_{wT_0}) and (μ_T, γ_{wT}) represent the pairs of viscosity and unit weight of water at temperatures T_0 and T, respectively.

Having determined the saturated hydraulic conductivity, the permeability is obtained as follows:

$$k = K_{sat}\frac{\mu}{\gamma_w} \quad (5\text{-}6)$$

The porosity n is related to the void ratio e as follows:

$$n = \frac{e}{1+e} \quad (5\text{-}7)$$

For calculation purposes one uses in Eq. (5-7) the average void ratio during a loading increment to yield the average porosity during that same increment.

This standard's methodology explains how to calculate the values of a_v, c_v, K_{sat}, k, and n for each step of incremental loading of the consolidation test. These variables change with changing effective stress instead of being constant values, as is the case when evaluating K_{sat} by falling-head, flexible-wall permeameter.

CHAPTER 6
DATA FOR EXAMPLE CONSOLIDATION TEST

Table 6-1 shows data for a consolidation test with seven load increments. This data set is used in this standard methodology to exemplify its calculation steps.

The characteristics of the tested soil specimen are initial height of specimen H_0 = 3.1852 cm, mass of solids = 343.57 g, density of solids = 2.65 g/cm^3, volume of solids = 129.65 cm^3, cross-sectional area of the specimen A = 93.31 cm^2, height of solids H_s = 1.3890 cm, initial void ratio (at the beginning of the test) e_0 = 1.292, and preconsolidation stress σ_p' = 80 kPa. The preconsolidation stress is the largest vertical effective stress that the soil specimen has been subjected to in situ.

6.1 THE AXIAL STRAIN VERSUS VERTICAL EFFECTIVE STRESS AND VOID RATIO VERSUS EFFECTIVE STRESS CURVES

Fig. 6-1 depicts the axial strain (ε_v) versus effective stress (σ_v') curve corresponding to the data in Table 6-1. The axial strain is calculated from the deformation readings listed in Table 6-1 according to Eqs. (3-1) and (3-2). The curve depicted in Fig. 6-1 changes slope at a preconsolidation vertical effective stress of about 100 kPa. This is the largest vertical effective stress that the soil specimen experienced in the field. The part of the curve corresponding to a vertical effective stress less than (more than) 100 kPa is the recompression (virgin) portion of the curve.

Fig. 6-2 shows the void ratio versus vertical effective stress curve for the example soil specimen. The void ratio is calculated from the deformation readings listed in Table 6-1 by means of Eq. (3-3).

The axial strain versus vertical effective stress and the void ratio versus vertical effective stress curves shown in Figs. 6-1 and 6-2 are for illustrative purposes only. They do not enter in the calculation of K_{sat}, k, and n but are used in the calculation of primary consolidation settlement (Loáiciga 2013).

Table 6-1. Consolidation Settlement Readings for a Seven-Increment Test

		Deformation readings (in 10^{-3} cm)						
		Loading increment (kPa)						
Elapsed time (min.)	Elapsed time ($\sqrt{}$min.)	12.26–24.52	24.52–49.04	49.04–98.07	98.07–196.14	196.14–392.28	392.28–784.56	784.56–1569.1
0.00	0.00	0.00	11.94	37.85	97.79	341.12	559.05	737.11
0.25	0.50	2.29	16.00	45.47	107.19	353.57	572.52	752.35
1.00	1.00	3.56	19.05	49.78	117.86	364.49	584.71	766.32
2.25	1.50	4.83	20.83	54.36	128.52	378.21	598.17	780.03
4.00	2.00	5.84	23.37	59.18	140.97	390.91	610.87	794.77
6.25	2.50	6.60	25.65	63.75	153.42	405.13	624.84	807.72
9.00	3.00	7.62	27.43	67.82	165.86	418.08	637.54	821.44
12.25	3.50	8.13	28.96	71.37	178.05	431.80	649.73	831.60
16.00	4.00	8.64	30.23	73.66	189.23	444.50	660.40	841.25
20.25	4.50	8.89	31.24	75.44	202.95	457.20	670.81	849.38
25.00	5.00	9.14	32.00	76.96	210.82	468.38	678.43	854.96
30.25	5.50	9.40	32.51	78.74	222.00	479.30	685.55	
36.00	6.00	9.65	33.02	79.76	230.12	487.93	692.40	863.60
42.25	6.50	9.91	33.78		238.25	495.55	698.00	
60.00	7.75	10.16	34.04	82.80	255.52	511.30	705.10	872.24
100.00	10.00	10.41	34.80	85.09	280.16	527.81	712.47	880.11
200.00	14.14	10.67	35.81	88.90	305.56	538.73	718.82	884.94
400.00	20.00	11.18	36.58	92.71	319.53	546.10	725.17	889.76
1440.00	37.95	11.94	37.85	97.79	341.12	559.05	737.11	897.64

Source: Adapted from Taylor (1948).

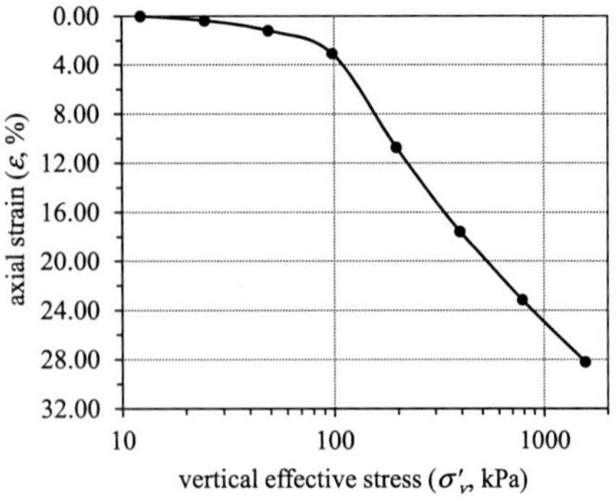

FIGURE 6-1. Axial Strain Versus Vertical Effective Stress for the Example Soil Specimen

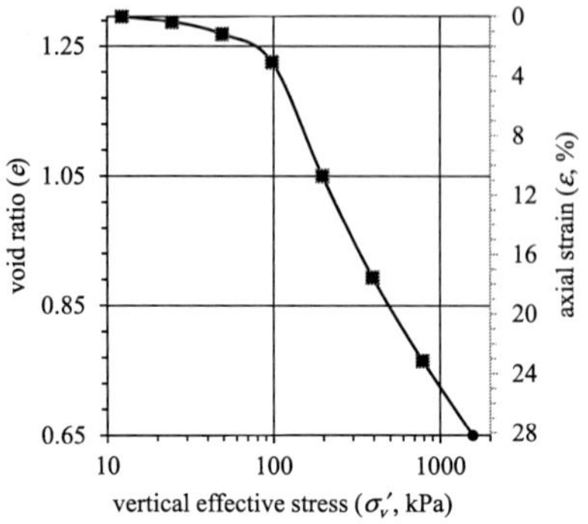

FIGURE 6-2. Void Ratio Versus Vertical Effective Stress for the Example Soil Specimen

CHAPTER 7
CALCULATION METHODOLOGY FOR VARIABLES K_{sat}, k, AND n OF FINE-GRAINED SOILS

This chapter presents methodology for the calculation of the saturated hydraulic conductivity (K_{sat}), permeability (k), and porosity (n) of fine-grained soils subjected to increasing vertical effective stresses. The exposition of the methodology is made concurrently with the presentation of a numerical example for clarity. The methodology of this standard guideline contains the following steps: (1) calculation of the coefficient of compressibility (a_v) as a function of the vertical effective stress, (2) calculation of the coefficient of consolidation (c_v) as a function of the vertical effective stress, (3) calculation of the saturated hydraulic conductivity (K_{sat}) as a function of the vertical effective stress, (4) calculation of the permeability (k) as a function of the vertical effective stress, and (5) calculation of the porosity (n) as a function of the vertical effective stress. The following subsections explain each of the five steps.

7.1 CALCULATION OF THE COMPRESSIBILITY COEFFICIENT (a_v)

Table 7-1 shows the deformation readings, vertical effective stress, calculated axial strain, and calculated void ratio for the example data set. The deformation readings and axial (vertical effective) stresses are taken from Table 6-1.

The coefficient of compressibility (a_v) is calculated with the following formula. See Eq. (3-4).

$$a_v = \frac{e_i - e_f}{\sigma'_{vf} - \sigma'_{vi}} \tag{7-1}$$

in which σ'_{vi} is the initial vertical effective stress of a load increment, e_i is the void ratio corresponding to the stress σ'_{vi}, σ'_{vf} equals the final vertical effective stress of a load increment, and e_f is the void ratio corresponding to the vertical effective stress σ'_{vf}. Notice that $e_f < e_i$ and $\sigma'_{vf} > \sigma'_{vi}$. The coefficient of compressibility is calculated for each load increment of the consolidation test. The results are listed in Table 7-2.

Fig. 7-1 portrays a graph of the coefficient of compressibility as a function of the average vertical effective stress for each load increment.

The coefficient of compression for the example data is seen in Fig. 7-1 to increase initially with increasing vertical effective stress until reaching a maximum value. Thereafter, it decreases monotonically.

7.2 CALCULATION OF THE COEFFICIENT OF CONSOLIDATION (c_v)

The calculation of the coefficient of consolidation as recommended in this standard guideline is carried out with Taylor's (1948) square-root-of-time fitting method. The latter method is illustrated in Fig. 7-2 in which the load increment from 392.32 to 784.56 kPa was chosen for specificity. The deformation data for the corresponding load increment are plotted as a function of the square root of the elapsed time since the beginning of the load increment. This produces the data line in Fig. 7-2. A straight line is traced passing through the initial, straight portion of the data line as shown in Fig. 7-2 and extended to intersect the horizontal axis at point A. The intersection value of the traced line (at point A) is multiplied by 1.15 to produce the value at point B in Fig. 7-2. Next, a straight line is drawn from point B to the origin of the data line. This line intersects the data line at point C. A vertical line is drawn from point C to the horizontal axis. The intersection point with the horizontal axis defines the square root of t_{90} (the square root of t_{90} is about 6.2 in Fig. 7-2); t_{90} is the time that it takes to achieve 90% of the primary consolidation of the test specimen subjected to the load increment (from 392.32 to 784.56 kPa in Fig. 7-2). In addition, t_{90} = 38.4 min in the example of Fig. 7-2.

The coefficient of consolidation is calculated as follows:

$$c_v = \frac{0.848 \cdot \left(\frac{H_i + H_f}{2}\right)^2}{t_{0.90}} \tag{7-2}$$

in which H_i is the height of the soil specimen when the applied stress equals the initial stress of the load increment (392.32 kPa in the example of Fig. 7-2), and H_f equals the height of the soil specimen corresponding to the final stress of the load increment (784.56 kPa in the example of Fig. 7-2). In addition, $t_{0.90}$ is calculated by squaring $\sqrt{t_{90}}$ read from the deformation versus the square root of the time graph.

The square-root-of-time fitting method is applied to each load increment of the consolidation test yielding as many calculated coefficients of consolidation as there are load increments. Table 7-3 lists the results of the calculated coefficient of consolidation as a function of vertical effective stress.

Fig. 7-3 graphs the coefficient of consolidation versus vertical effective stress data listed in Table 7-3.

The function of the coefficient of consolidation as a function of vertical effective stress for the sample data is complex. It starts with the largest value at low axial stress and decreases until an axial stress of about 200 kPa is reached. Thereafter, it increases with increasing axial stress. The increase of the coefficient of consolidation for vertical effective stress larger than 200 kPa might be related to the fact that the virgin portion of the consolidation curve does not straighten up with vertical effective stress larger than 200 kPa but rather has a slight upward curvature, as seen in Fig. 6-2.

Table 7-1. Calculated Axial Strain and Void Ratio; Axial Stress and Deformation from Table 6-1

Axial stress, kPa (experimental)	Deformation, cm (measured)	Specimen height, cm Eq. (3.1)	Axial strain (%) Eq. (3.2)	Void ratio (e) Eq. (3.3)
12.26	0.0000	$H_0 = 3.1852$	0	$e_0 = 1.292$
24.52	0.0119	3.1733	0.375	1.283
49.04	0.0378	3.1474	1.188	1.265
98.07	0.0978	3.0874	3.070	1.222
196.14	0.3411	2.8441	10.710	1.047
392.28	0.5591	2.6261	17.552	0.890
784.56	0.7371	2.4481	23.142	0.762
1569.1	0.8976	2.2876	28.181	0.646

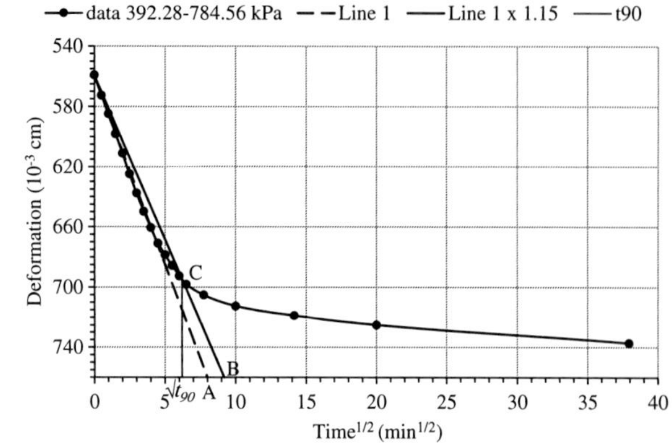

FIGURE 7-2. Taylor's Square-Root-of-Time Method to Calculate the Coefficient of Consolidation

Table 7-2. Calculated Coefficient of Compressibility for Each Load Increment; Axial Stress and Void Ratio from Table 7-1

Axial stress, kPa	Average axial stress, kPa	Void ratio (e) Eq. (3-3)	Coefficient compressibility, $a_v \times 10^4$, 1/kPa Eq. (7-1)
12.26		1.292	
	18.39		7.01
24.52		1.283	
	36.78		7.60
49.04		1.265	
	73.555		8.80
98.07		1.222	
	147.11		17.85
196.14		1.047	
	294.21		8.00
392.28		0.890	
	588.42		3.27
784.56		0.762	
	1176.8		1.47
1569.1		0.646	

Table 7-3. Coefficient of Consolidation Results

Axial stress, kPa	Average axial stress, kPa	Coefficient consolidation (c_v), mm²/s Eq. (7-2)
12.26		
	18.39	1.23
24.52		
	36.78	1.15
49.04		
	73.555	0.74
98.07		
	147.11	0.19
196.14		
	294.21	0.17
392.28		
	588.42	0.25
784.56		
	1176.8	0.36
1569.1		

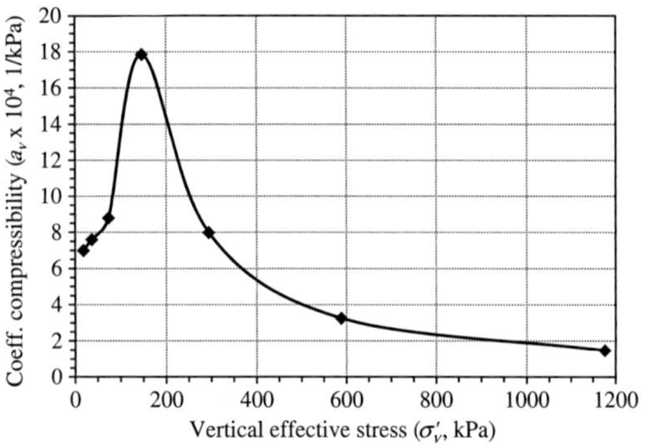

FIGURE 7-1. Coefficient of Compressibility as a Function of the Vertical Effective Stress

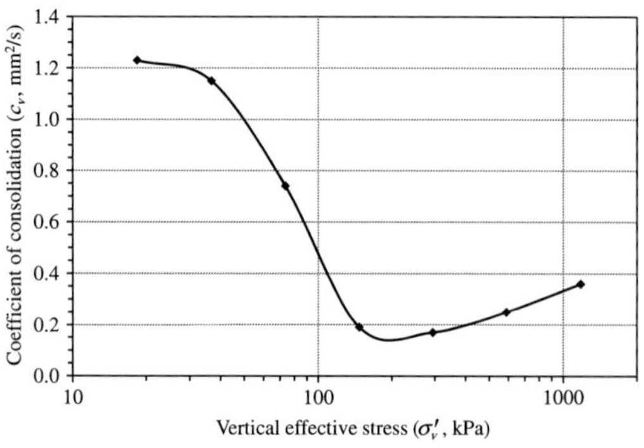

FIGURE 7-3. Graph of the Coefficient of Consolidation Versus Vertical Effective Stress; Data from Table 7-3

7.3 CALCULATION OF THE SATURATED HYDRAULIC CONDUCTIVITY (K_{sat}).

The saturated hydraulic conductivity in the r-th loading increment (K_{sat}) is calculated with the following formula. See Eq. (5-3).

$$K_{sat} = \frac{c_v \gamma_w a_v}{1 + e_i} \tag{7-3}$$

in which a_v and c_v represent the calculated coefficient of compressibility and coefficient of consolidation for the r-th loading increment, respectively, and e_i denotes the void ratio at the beginning of the r-th loading increment. Table 7-4 lists the calculated values of K_{sat} for the example data.

Fig. 7-4 graphs the variation of saturated hydraulic conductivity as a function of the vertical effective stress.

Notice in Fig. 7-4 that the hydraulic conductivity decreases with increasing vertical effective stress except for low levels of vertical effective stress, where a slight increase in K_{sat} is observed. The slight increase in K_{sat} at low vertical effective stress is counterintuitive, because the pore volume is reduced with increasing vertical stress, which one would expect would diminish the capacity of a soil to transmit water. A plausible explanation of the observed initial increase in K_{sat} is that as the soil specimen is compressed, entrapped air escapes, temporarily increasing the flow of water out of the sample. Thus, the initial increase in K_{sat} seen in Fig. 7-4 may be due to experimental limitations, that is, to the difficulty of completely de-airing the soil specimen prior to consolidation testing, rather than to some other anomalous physical phenomenon. The decline of K_{sat} with increasing vertical effective stress takes place once any air present in the sample has been removed.

There is a one-order-of-magnitude reduction in K_{sat} according to the data presented in Fig. 7-4 as the vertical effective stress increases from its lowest to its highest level in the example data. For the purpose of groundwater flow modeling in compressible strata, the data shown in Fig. 7-4 imply that it is necessary to modify the saturated hydraulic conductivity as the vertical effective stress increases. This must be done for each fine-grained, compressible stratum whose hydraulic conductivity is affected by increasing vertical effective stress.

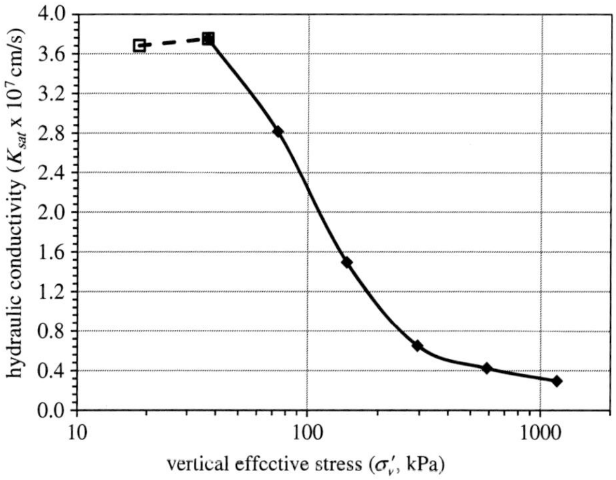

FIGURE 7-4. The Saturated Hydraulic Conductivity as a Function of the Vertical Effective Stress

7.4 CALCULATION OF THE PERMEABILITY (k)

The permeability is calculated in each loading increment with the following formula. See Eq. (5-6).

$$k = K_{sat} \frac{\mu_T}{\gamma_{wT}} \tag{7-4}$$

The symbols μ_T and γ_{wT} denote the dynamic viscosity and the unit weight of water at a temperature T used in the consolidation test, respectively. In the example data, the water temperature equals 20°C. Thus, $\mu_T = 0.010050\,\text{g}/(\text{cm}\cdot\text{s})$ and $\gamma_{wT} = 9.79\,\text{kN}/\text{m}^3$. Eq. (7-4) was applied to the saturated hydraulic conductivity values reported in Table 7-4 to produce the permeability listed in Table 7-5.

Fig. 7-5 depicts the calculated permeability as a function of the axial stress. The graph of the permeability is a scaled version of that pertaining to the saturated hydraulic conductivity shown in Fig. 7-4.

Table 7-4. Calculated Values of the Saturated Hydraulic Conductivity

Axial stress, kPa	Average axial stress, kPa	Void ratio (e) Eq. (3-3)	a_v x 10^4 1/kPa Eq. (7-1)	c_v mm^2/s Eq. (7-2)	K_{sat} x 10^7 cm/s Eq. (7-3)
12.26		1.292			
	18.39		7.01	1.23	3.68
24.52		1.283			
	36.78		7.60	1.15	3.75
49.04		1.265			
	73.555		8.80	0.74	2.81
98.07		1.222			
	147.11		17.85	0.19	1.49
196.14		1.047			
	294.21		8.00	0.17	0.65
392.28		0.890			
	588.42		3.27	0.25	0.42
784.56		0.762			
	1176.8		1.47	0.36	0.29
1569.1		0.646			

Table 7-5. Calculated Permeability Values (1 microdarcy = 9.87 × 10^−15 cm^2)

Axial stress, kPa	Average axial stress, kPa	K_{sat} x 10^7 cm/s Eq. (7.3)	Permeability (k), microdarcy Eq. (7.4)
12.26			
	18.39	3.68	383
24.52			
	36.78	3.75	390
49.04			
	73.555	2.81	293
98.07			
	147.11	1.49	155
196.14			
	294.21	0.65	68
392.28			
	588.42	0.42	44
784.56			
	1176.8	0.29	31
1569.1			

Table 7-6. Calculation of the Porosity for Several Loading Increments

Axial stress, kPa	Average axial stress, kPa	Void ratio (e) Eq. (3.3)	Average void ratio ($\bar{e}$) Eq. (7.5)	Porosity (n) Eq. (7.6)
12.26		1.292		
	18.39		1.29	0.563
24.52		1.283		
	36.78		1.27	0.560
49.04		1.265		
	73.555		1.24	0.554
98.07		1.222		
	147.11		1.13	0.531
196.14		1.047		
	294.21		0.97	0.492
392.28		0.890		
	588.42		0.83	0.452
784.56		0.762		
	1176.8		0.70	0.413
1569.1		0.646		

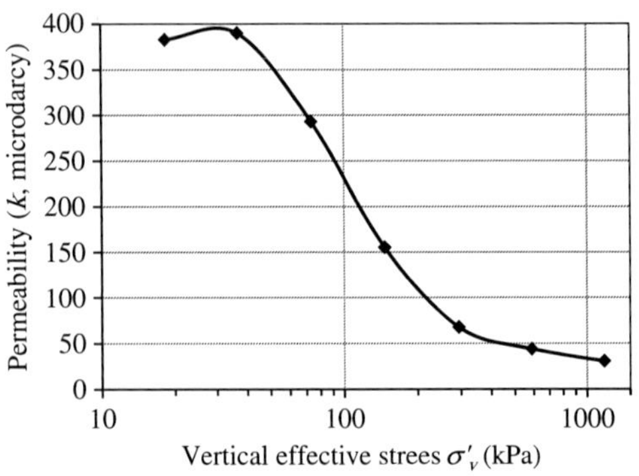

FIGURE 7-5. Plot of the Permeability Versus the Vertical Effective Stress

7.5 CALCULATION OF THE POROSITY (n)

The porosity of a soil specimen decreases with increasing vertical effective stress. For calculation purposes, the porosity during the r-th loading increment is obtained by applying Eq. (5-7) to the average void ratio $\bar{e}$ during the r-th loading increment, where the average void ratio is calculated as follows:

$$\bar{e} = \frac{e_i + e_f}{2} \tag{7-5}$$

in which e_i and e_f denote the void ratios of a soil specimen at the beginning and end of a loading increment, respectively. Therefore,

$$n = \frac{\bar{e}}{1 + \bar{e}} \tag{7-6}$$

Eq. (7-6) was applied to the deformation versus stress data listed in Table 6-1 to obtain the results shown in Table 7-6.

Fig. 7-6 shows the plot of the porosity as a function of the vertical effective stress. The graph in Fig. 7-6 exhibits a monotonically decreasing porosity with increasing vertical effective stress.

Sections 7.1 to 7.5 in this chapter have detailed the methodology to calculate the saturated hydraulic conductivity, permeability, and porosity of fine-grained soils subjected to increasing vertical effective stress and provided calculation examples. The next chapter reviews an alternative method that has been reported to calculate the K_{sat} with a flexible-wall permeameter.

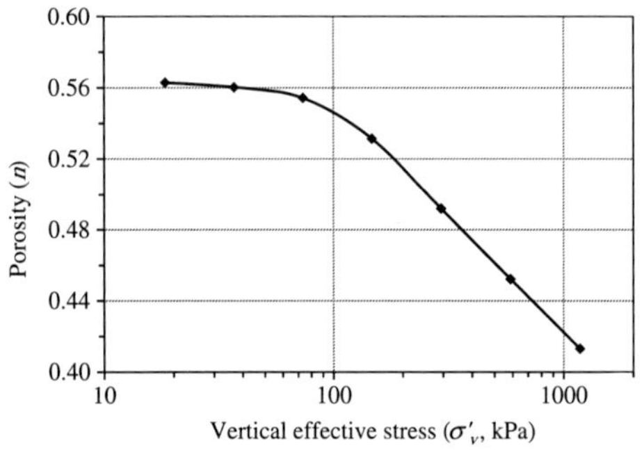

FIGURE 7-6. Graph of the Porosity Versus the Vertical Effective Stress

CHAPTER 8
ANOTHER APPROACH: FLEXIBLE-WALL PERMEAMETER

The saturated hydraulic conductivity of fined-grained soils is commonly so low that it renders field slug tests or pumping tests impractical for its determination. An alternative method for the laboratory determination of the saturated hydraulic conductivity is the use of the flexible-wall permeameter as described in the ASTM test method D5084-10 with hydraulic systems A, B, C, D, E, and F for soil specimens with saturated hydraulic conductivity less than 1×10^{-4} cm/s (ASTM 2010). Vertical stresses can be applied to the soil specimen placed in the cell of the flexible-wall permeameter to simulate the in situ vertical effective stress. Experimental difficulties arise, however, due to leakage through the permeameter's cell containing the soil specimen and shearing failure of the specimen at relatively high compressive stress. For this reason, the consolidometer test-method described and exemplified in this standard guideline is recommended for the determination of the saturated hydraulic conductivity, permeability, and porosity under variable vertical stresses for low-permeability, fine-grained soils. Test method D5084-10 with hydraulic system B is briefly described for the sake of completeness. This method uses a falling-head permeameter with constant tailwater elevation in the determination of the saturated hydraulic conductivity of low-permeability soils. This is the most commonly used version of test method D5084-10 (see, e.g., Cedergreen 1989 for a review of test methods for the determination of the saturated hydraulic conductivity).

Fig. 8-1 shows a diagram of a flexible-wall permeameter featuring a hydraulic system with falling head and constant tailwater elevation (ASTM test method D5084-10 variant B).

The cross-sectional areas of the water-influent line and the permeameter's cell containing the soil specimen are denoted by a and A, respectively, in Fig. 8-1. The length of the soil specimen in the permeameter's cell is denoted by L with h_0 and h_f as the head losses through the soil specimen at the initiation of the test and the end of the test, respectively. These hydraulic head losses are defined with respect to the constant water level of the tailwater, as shown in Fig. 8-1. They are usually enlarged by pressuring the influent line. Large hydraulic head losses may cause leaks in the permeameter's cell. The time elapsing between the initiation of the test and the end of the test is represented by Δt. The saturated hydraulic conductivity for the applied axial stress is derived by equating the volume of water displaced in the influent line as the hydraulic head declines during the test to the volume of water that moves through the soil specimen during the test duration calculated with Darcy's law. The resulting formula for the saturated hydraulic conductivity follows (see Cedergreen 1989 or ASTM 2010).

$$K_{sat} = \frac{a\,L}{A\,\Delta t} \ln\left(\frac{h_0}{h_f}\right) \tag{8-1}$$

Eq. (8-1) calculates K_{sat} from hydraulic principles. Eq. (7-3), conversely, measures K_{sat} using deformation data directly over a wide range of vertical effective stress.

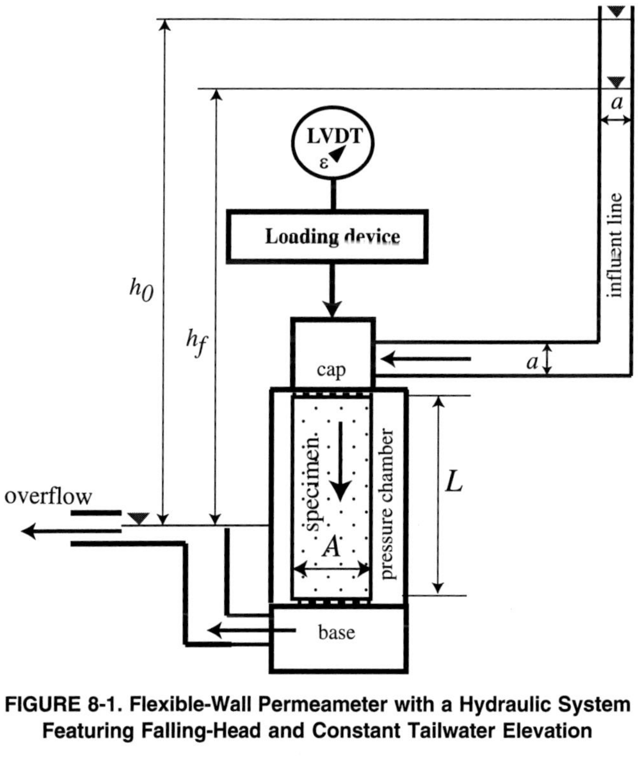

FIGURE 8-1. Flexible-Wall Permeameter with a Hydraulic System Featuring Falling-Head and Constant Tailwater Elevation

CHAPTER 9
SECONDARY CONSOLIDATION SETTLEMENT: K_{sat} AT CONSTANT STRESS

Secondary consolidation settlement is the vertical deformation of soil that follows after the end of primary consolidation settlement in saturated, fine-grained soils. Unlike primary consolidation settlement, secondary consolidation settlement takes place at constant vertical effective stress. Secondary consolidation settlement occurs once the excess pore pressure induced by the external loads has vanished in the case of applied external loads. In the case of groundwater extraction, secondary consolidation settlement occurs once, and if, the hydraulic head produced by groundwater pumping reaches steady state. The cause of continued deformation of soils under constant effective stress is not well understood. It has been hypothesized that it is caused by microscale changes in the structure of clay particles (Holtz et al. 2011). Regardless of its origin, secondary consolidation settlement reduces the void ratio of soils beyond the effects of primary consolidation settlement, and, therefore, diminishes their porosity, permeability, and hydraulic conductivity.

Fig. 9-1 depicts the reduction in void ratio as a function of time during the loading increment from 392.28 to 784.56 kPa corresponding to the data shown in Table 6-1. Symbols e_i, ε_i, H_i, and K_i represent the void ratio, axial strain, soil specimen's height, and saturated hydraulic conductivity at the beginning of the load increment, respectively, whereas e_f, ε_i, H_f, and K_i are respectively the void ratio, axial strain, specimen's height, and saturated hydraulic conductivity at the end of the loading increment.

Fig. 9-1 depicts the customary procedure used to empirically separate the primary and secondary consolidation phases. Tangent lines tang 1 and tang 2 are extended through the straight parts of the primary and secondary consolidation data, respectively. Their point of intersection is defined to mark the time when primary consolidation ends (t_P) and the onset of secondary consolidation, shown in Fig. 9-1 to be approximately 43 min for the tested soil specimen. The time that it takes in the field to reach the end of primary consolidation is much longer. It can be months or decades, depending on the drainage properties of a soil mass. The slope of the tangent line tang 2, called the secondary compression index (C_α), is used to predict the secondary consolidation settlement that might occur in the field when a similar increment in vertical loading occurs as that applied in the laboratory test. Incidentally, the secondary compression index calculated from Fig. 9-1 equals 0.015.

Assume that a soil stratum is subjected to an increment of vertical effective stress ranging from 392.28 to 784.56 kPa and that the time to complete primary consolidation settlement equals t_P years. Let H_i and e_i be the height and void ratio of the soil stratum just before the application of the considered increment of vertical effective stress, and the secondary compression index be denoted by C_α. The secondary consolidation settlement (s_s) that would occur by time $t > t_P$ is given by the following formula:

$$s_s(t) = \frac{C_\alpha}{1+e_i} H_i \log_{10} \frac{t}{t_P} \quad t > t_P \tag{9-1}$$

Secondary consolidation settlement can also be calculated using the modified secondary compression index ($C_{\alpha\varepsilon}$), which is obtained as the slope of the line tang 2 in Fig. 9-1 when the vertical strain replaces the void ratio as the variable on the vertical axis. In this case,

$$s_s(t) = C_{\alpha\varepsilon} H_i \log_{10} \frac{t}{t_P} \quad t > t_P \tag{9-2}$$

Notice that $C_{\alpha\varepsilon} = C_\alpha/(1+e_i)$. The secondary consolidation settlement calculated with Eqs. (9-1) or (9-2) equals the reduction in thickness of a soil stratum after primary consolidation is completed. The volume of water drained during a period of secondary consolidation affecting an area (A) of stratum evenly equals $s_s \cdot A$. There is no explicit information that arises from this analysis insofar as the variation of the saturated hydraulic conductivity is concerned.

The void ratio decreases as a result of primary and secondary consolidation. The axial strain during secondary consolidation equals $\varepsilon_s = s_s/H_i$. The axial strain produced during primary consolidation (ε_P) can be obtained with standard methodology (see, e.g., Holtz et al. 2011). Thus, the total axial strain is the sum

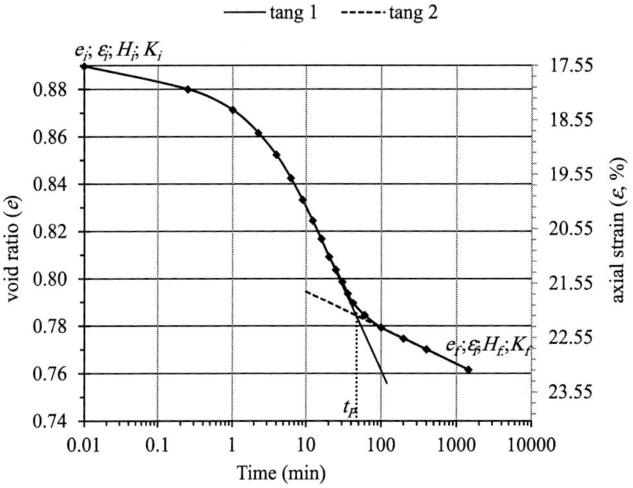

FIGURE 9-1. Reduction in Void Ratio as a Function of Time for the Loading Increment from 392.28 to 784.56 kPa

of the axial strains that occur during primary and secondary consolidation, $\varepsilon = \varepsilon_P + \varepsilon_s$. The change in void ratio equals

$$\Delta e = (\varepsilon_s + \varepsilon_P) \cdot (1 + e_i) \tag{9-3}$$

The void ratio at time $t \geq t_P$ is then

$$e = e_i - \Delta e \tag{9-4}$$

One approach to predict the saturated hydraulic conductivity during secondary consolidation is to use the following formula proposed by Smarasinghe et al. (1982) that relates the hydraulic conductivity to the void ratio of clays:

$$K_{sat} = C \frac{e^m}{(1 + e)} \tag{9-5}$$

where C and m are coefficients estimable from measurements of K_{sat} and e. The void ratio e in Eq. (9-5) is given by Eq. (9-4). Eq. (9-5) calculates the saturated hydraulic conductivity as a function of time during secondary consolidation. The coefficients C and m are calculated from the deformation measurements during the loading increment of interest. Referring to the nomenclature introduced in Fig. 9-1, it can be shown that the coefficients are given by the following formulas:

$$m = \frac{1}{ln\left(\frac{e_i}{e_f}\right)} \cdot ln\left(\frac{K_i}{K_f} \cdot \frac{1 + e_i}{1 + e_f}\right) \tag{9-6}$$

$$C = K_r \frac{(1 + e_r)}{e_r^m} \qquad r = i \, \text{or} \, f \tag{9-7}$$

The coefficients C and m are substituted in Eq. (9-5) to yield the time-dependent saturated hydraulic conductivity during secondary compression for the loading interval of interest.

Last, the porosity varies during secondary porosity according to its relation with the void ratio given by Eq. (5-7). The void ratio, in turn, is given by Eq. (9-4).

This standard's data showing a decline of K_{sat} with increasing vertical effective stress (see Fig. 7-4) have implications for groundwater modeling. Specifically, it is necessary to modify the saturated hydraulic conductivity as the void ratio decreases. This must be done for each fine-grained, compressible stratum whose hydraulic conductivity is changed by decreasing void ratio. One word of caution in this regard is that the consolidation of compressible clays is an inelastic phenomenon, that is, upon relief of vertical effective stress, the clay rebounds following a path different to that followed during consolidation and recovers only a fraction of its lost thickness. Fortunately, 1D consolidation tests can execute loading and unloading cycles to provide all the necessary data needed in groundwater models.

CHAPTER 10
REFERENCES

ASCE. (2008a). "Standard guideline for fitting saturated hydraulic conductivity using probability density function." *ASCE/EWRI 50-08*, Reston, VA.

ASCE. (2008b). "Standard guideline for calculating the effective saturated hydraulic conductivity." *ASCE/EWRI 51-08*, Reston, VA.

ASCE. (2010). "Standard guideline for the geostatistical estimation and block-averaging of homogeneous and isotropic saturated hydraulic conductivity." *ASCE/EWRI 54-10*, Reston, VA.

ASTM. (2005). "Standard guide for selection of soil and rock sampling devices used with drill rigs for environmental investigations." *ASTM D6169*, West Conshohocken, PA.

ASTM. (2009). "Standard test method for laboratory determination of pulse velocities and ultrasonic elastic constants of rock." *ASTM D2845-08*, West Conshohocken, PA.

ASTM. (2010). "Standard test methods for measurement of hydraulic conductivity of saturated porous materials using a flexible wall permeameter." *ASTM D5084-10*, West Conshohocken, PA.

ASTM. (2011a). "Standard terminology relating to soil, rock, and contained fluids." *ASTM D653–11*, West Conshohocken, PA.

ASTM. (2011b). "Standard test methods for one-dimensional consolidation properties of soils using incremental loading." *ASTM D2435/D2435M-11*, West Conshohocken, PA.

Bowles, J. E. (1996). *Foundation analysis and design*, McGraw-Hill, New York.

Carslaw, H. S., and Jaeger, J. C. (1986). *Conduction of heat in solids*, 2nd Ed., Clarendon Press, Oxford.

Cedergreen, H. R. (1989). *Seepage, drainage, and flow nets*, Wiley, New York.

Das, B. (2015). *Principles of foundation engineering*, Cengage Learning, Boston.

Duncan, J. M., and Chang, C. Y. (1970). "Nonlinear analysis of stress and strain in soils." *J. Soil. Mech. Found. Div.*, 96(SM5), 1629–1653.

Fetter, C. W. (2001). *Applied hydrogeology*, Prentice Hall, Upper Saddle River, NJ.

Gambolati, G., Teatini, P., Baú, D., and Ferronato, M. (2000). "Importance of poroelastic coupling in dynamically active aquifers of the Po River basin, Italy." *Water Resourc. Res.*, 36(9), 2443–2459.

Griffiths, D. V., and Marquez, R. M. (2007). "Three-dimensional slope stability analysis by elasto-plastic elements." *Géotechnique*, 57(6), 537–546.

Holtz, R. B., Kovacs, W. D., and Sheahan, T. C. (2011). *An introduction to geotechnical engineering*, Pearson, Upper Saddle River, NJ.

Ingebritsen, S., Sanford, W., and Neuzil, C. (2006). *Groundwater and geologic processes*, Cambridge University Press, Cambridge, U.K.

Kang, D. H., and Li, J. (2015). "Comparing simulations of land subsidence using a 3-D package NDIS and 1-D subsidence package SUB." *Floods, droughts, and ecosystems: Proc., World Environmental and Water Resources Congress*, ASCE Press, Reston, VA, 486–495.

Loáiciga, H. A. (2013). "Consolidation settlement in aquifers caused by pumping." *J. Geotech. Geoenviron. Eng.*, 139(7), 1191–1204.

Marcuson, W. F., III, and Franklin, A. G. (1979). "State of the art of undisturbed sampling of cohesionless soils." Miscellaneous Paper GL-79-16, Geotechnical Laboratory, U.S. Army Engineer Waterways Experimental Station, Vicksburg, MS.

Nur, A., and Byerlee, J. D. (1971). "An exact effective stress law for effective stress for elastic deformation of rock with fluid." *J. Geophys. Res.*, 76, 6414–6419.

Samarasinghe, A. M., Huang, Y. H., and Drnevich, V. P. (1982). "Permeability and consolidation of normally consolidated clays." *J. Geotech. Eng. Div.*, 108(6), 835–850.

Skempton, A. W., and Bjerrum, L. (1957). "A contribution to the settlement analysis of foundations on clay." *Géotechnique*, 7, 168–178.

Taylor, D. W. (1948). *Fundamentals of soil mechanics*, Wiley, New York.

Terzaghi, K. (1925). "Principles of soil mechanics." *Eng. News Rec.*, 95, 25–27.

Timoshenko, S. P., and Goodier, J. N. (1970). *Theory of elasticity*, McGraw-Hill, New York.

USACE (U.S. Army Corps of Engineers). (1980). "Engineer manual." EM 1110-2-1906, Laboratory Soils Testing. Dept. of the Army, Office of the Chief of Engineers, Washington, DC.

Weinberger, H. F. (1965). *A first course in partial differential equations*, Wiley, New York.

Zektser, I. S., Loáiciga, H. A., and Wolf, J. (2005). "Environmental impacts of groundwater overdraft: Selected case studies in the southwestern United States." *J. Environ. Geol.*, 47(3), 396–404.

CHAPTER 11
NOTATION

a : cross-sectional area of the influent line of the flexible-wall permeameter (length squared)
A : cross-sectional area of the soil specimen in the cell of the flexible-wall permeameter or in the consolidometer (length squared)
c_v : coefficient of consolidation (length squared over time)
C : coefficient in the formula relating the saturated hydraulic conductivity to the void ratio
d_r : steady-state drawdown at a distance, r, for an extraction well (length)
e : void ratio
e_0 : void ratio of soil specimen before it is subject to axial compression in a consolidometer
e_i : void ratio of a soil specimen at the beginning of a loading increment
e_f : void ratio of a soil specimen at the end of a loading increment
e_t : void ratio of a soil specimen after a time, t, of compression in the consolidometer
$\bar{e}$: average void ratio during a loading increment
E : modulus of elasticity (kPa)
E_d : constrained modulus (kPa)
h_0 : initial hydraulic head loss through the soil specimen in the flexible-wall permeameter (length)
h_f : final hydraulic head loss through the soil specimen in the flexible-wall permeameter (length)
H_0 : initial height of a soil specimen before being subjected to compression in the consolidometer, or, the initial thickness of a stratum of soil (length)
H_i : initial height of a soil specimen at the beginning of a loading increment (length)
H_f : height of a soil specimen at the end of a soil increment (length)
H_t : height of a soil specimen after a time, t, of compression in the consolidometer (length)
H_1 : initial height of a soil specimen at the beginning of a loading increment (length)
H_i : initial height of a soil specimen at the beginning of a loading increment (length)
H_f : height of a soil specimen at the end of a loading increment (length)
k : permeability (length squared)
K_{sat} : saturated hydraulic conductivity (length over time)
K_0 : at-rest coefficient of lateral earth pressure
L : length of the soil specimen in the flexible-wall permeameter
m : parameter in the formula relating the saturated hydraulic conductivity to the void ratio
n : porosity of a soil specimen
P : vertical axial load in the consolidometer test (kN)
q : vertical stress at the bottom of an embankment (kPa)
r : distance from the center of an extraction well to a point where drawdown is calculated (length)
R_0 : deformation reading at the beginning of a consolidation test (length)
R_t : deformation reading at a time, t, since the beginning of a consolidation test (length)
$t_{0.90}$: the time that it takes to achieve 90% of the primary consolidation of the test specimen subjected to a load increment in the consolidation test (minutes)
T : temperature of water (°C)
T_0 : reference temperature of water (typically 20°C)
u : pore pressure (kPa)
u_e : excess pore pressure (kPa)
z : depth of a soil element below the ground surface (length)
α : the Biot coefficient given by $\alpha = 1 - (\alpha_s/\alpha_b)$
α_b : coefficient of compressibility of a soil's bulk matrix (solids plus pores) (1/kPa)
α_s : coefficient of compressibility of a soil's solids (1/kPa)
γ : unit weight of soil (kN/m^3)
γ_w : unit weight of water (kN/m^3)
γ_{wT} : unit weight of water at a temperature, T (kN/m^3)
γ_{wT_0} : unit weight of water at a temperature, T_0 (kN/m^3)
δ_r : land subsidence at the distance, r, from an extraction well (length)
ε : axial (vertical) strain in a soil specimen
ε_P : axial (vertical) strain in a soil specimen at the end of primary consolidation
ε_s : axial (vertical) strain in a soil specimen during secondary consolidation
ε_t : axial (vertical) strain in a soil specimen at time, t
μ : dynamic viscosity of water [g/(s cm)]
μ_T : dynamic viscosity of water at temperature T [g/(s cm)]
μ_{T_0} : dynamic viscosity of water at temperature, T_0 [g/(s cm)]
ν : Poisson's ratio
σ'_h : horizontal effective stress (kPa)
σ'_v : vertical effective stress (kPa)
σ'_{h0} : initial horizontal effective stress (kPa)
σ'_{v0} : initial vertical effective stress (kPa)
σ'_{vt} : vertical effective stress at time, t
ΔH_t : change in height of a soil specimen after a time, t, of undergoing compression in a consolidometer (length)
ΔH_1 : change in height of a soil specimen after one load increment in a consolidometer (length)
ΔH_2 : change in height of a soil specimen after two load increments in a consolidometer (length)
ΔH_t : reduction in the height of a soil specimen or soil stratum after a time, t, of compression (length)
$\Delta\sigma'_v$: change in vertical effective stress (kPa)
$\Delta\sigma'_h$: change in horizontal effective stress (kPa)

INDEX